AMAL HAMMAMI

Management and Security of a computer park

AMAL HAMMAMI

Management and Security of a computer park

ScienciaScripts

Imprint
Any brand names and product names mentioned in this book are subject to trademark, brand or patent protection and are trademarks or registered trademarks of their respective holders. The use of brand names, product names, common names, trade names, product descriptions etc. even without a particular marking in this work is in no way to be construed to mean that such names may be regarded as unrestricted in respect of trademark and brand protection legislation and could thus be used by anyone.

Cover image: www.ingimage.com

This book is a translation from the original published under ISBN 978-613-9-51158-7.

Publisher:
Sciencia Scripts
is a trademark of
Dodo Books Indian Ocean Ltd. and OmniScriptum S.R.L publishing group

120 High Road, East Finchley, London, N2 9ED, United Kingdom
Str. Armeneasca 28/1, office 1, Chisinau MD-2012, Republic of Moldova, Europe
Printed at: see last page
ISBN: 978-620-5-97930-3

Contents

GENERAL INTRODUCTION

Since the manual management of tasks in a system encounters difficulties due to the diversity and mass of information handled, IT was born as a vast domain without precise boundaries to simplify things and automate applications, no matter how complex they are. IT has become an essential management, organisation, production and communication tool for the company[6]. Today, with the impetus of new technologies such as the Internet and multimedia, we are witnessing a real change in the ways of exchanging information in the classical administrative procedures. One of the advantages of IT is the possibility of network deployment. The company's network implements sensitive data, saves them, shares them in internal elements and sometimes communicates them to other companies or persons via external networks. [7] Computer networks thus increase the productivity and competitiveness of employees and offer industries new methods of marketing products and services. As a result, it becomes impossible to forego the benefits of information, to isolate, in other words, the network from the outside world and to remove the electronic and confidential nature of the data. 8] The sensitive data in the company's information system are therefore exposed to malicious acts, the nature and method of intrusion of which are constantly changing. In this context, faced with the development of computer maliciousness, both internal and external, security needs are increasingly strong. In order to build a robust and adapted security solution.

To ensure our *IT* Asset Management and Security solution, the working methodology will be presented in four chapters which are as follows

Firstly, in the first chapter we propose to present the general framework of our project and the different activities during this period, we will identify the problematic that will lead us to work on this project.

Secondly, in the second chapter, we will introduce the notion of computer security and clarify the different types of threats and possible attacks.

In the third chapter, we will propose an architecture dëtaillëe to the EAH firm's computer park while installing and configuring the necessary doniK'e servers.

In the last chapter, we will secure this architecture by applying a security strategy by integrating a set of IT tools such as firewalls, proxy servers, ssh servers, etc.

Finally, we close this work with a conclusion дёпёгак and perspectives.

CHAPTER 1

Project background

1. INTRODUCTION

In this chapter, we first present the needs that the application addresses and the value it brings, while critiquing existing solutions.

2. Analysis of the problem

With the evolution of the Internet, companies are increasingly opening their information systems to their partners or suppliers, so it is essential to know which company resources to protect and to control user rights and access to their information systems. The same applies when opening up the company's access to the Internet.

Generally speaking, IT security is about ensuring that an organisation's hardware or software resources are only used for their intended purpose. In this context,

3. Terms of reference

The *EAH* firm proposed us as a final study project the Management and Security of its IT park. During this internship, our work was articulated around several tasks such as:

J Installation of a linux server and application of a well studied strategy for the management and security of the data of the EAH firm.

J Configuring other PCs as clients to form the EAH computer pool.

J Router configurations to meet the needs of the practice.

4. Work planning

To dëvelop the solution of the sëcuritë of information systems. Our work has ële plaшйё as follows:

Table 1: Work planning

	FEBRUARY	MARCH	APRIL	MAY	JUNE
Computer Systems Training	x	x	x	x	x
General context of the project	x	x			
Information Strategy and Security in the Enterprise		x	x		
*Study of **the** archil ecl uze*			x	x	
*Application **of** Security to our Architecture*				x	x

5. Conclusion

In this introductory chapter, we have presented the дёпёгак framework of our project, with a study of the existing.

We ёШюиз in the following chapter the notion of computer sëcuritë while specifying the different types of threats and possible attacks.

CHAPTER 2

Information Strategy and Security in the Enterprise

1. INTRODUCTION

The growth of computer systems and the expansion of the number of interconnected networks make the issue of computer sëcuritë increasingly important. Computer security has nowadays become a major issue in the management of corporate networks as well as for the ever increasing number of individuals connecting to the Internet. The transmission of sensitive information and the desire to ensure its confidentiality has become a key issue in the implementation of computer networks.

Network and information security is defined as the ability of a network or information system to withstand accidental events or malicious actions. It is also defined as the ability of a system to protect objects while respecting their confidentiality, integrity and availability. Security incidents can be grouped as follows:

J Communication can be intercepted and data can be copied or modified.

Unauthorised access to a computer or computer network is usually a malicious attempt to copy, modify or destroy data. Disruptive attacks on the internet have become quite common.

J Malicious software, such as viruses, home in the middle, spyware...can erase and modify data.

Misrepresentation of persons or entities can cause substantial damage, e.g. customers may download malicious software from a website posing as a trusted source, contracts may be repudiated, confidential information may be sent to the wrong recipient. [3]

Many security problems are due to unforeseen or unintentional events such as natural disasters (floods, storms, earthquakes), hardware and software failures, human error.

2. The purpose of IT security

2.1 Definition

Computer security is the set of technical, organisational, legal and human means necessary and put in place to preserve, restore, and guarantee the sëcuritë of an organisation's information system resources (hardware or software).

2.2 Description

Computer security is composed of three steps:

> Risk analysis
> Security policy
> Security techniques

2.2.1 Risk analysis

No company can do without information technology, which is why it is necessary to ensure its security and to protect itself against attacks. Since we can only protect ourselves effectively against known risks, it is important to measure them in terms of the probability or frequency of their occurrence and their possible effects.

2.2.2 Security policy

In the light of the results of the risk analysis, the security policy :

> Defines the framework for the use of information system resources;

> Identifies the security techniques to be used in the various

services of the organisation ;

> Raise awareness of computer security among users.

2.2.3 Security techniques

They ensure availability (services and information must be accessible to authorised persons when they need them and within the required timeframe), integrity (services and information can only be modified by authorised persons), and confidentiality (information is only accessible to those who have the right to the data transmitted).

Techniques for securing a system include:

> Vulnerability audit ;

> Data security: encryption, authentication, access control ;

> Network security: firewall, proxy ;

> Monitoring security information ;

2.3 Objective

The objective of computer security is the implementation of protection mechanisms to ensure the following properties [3]:

Confidentiality: Confidentiality is the maintenance of the secrecy of information and flows, of transactions. It is the protection of resources from unauthorised disclosure. Confidentiality can be achieved by implementing access control or encryption mechanisms. Data encryption (or cryptography), helps to ensure the confidentiality of information during transmission or storage.

In general, ensuring confidentiality means ensuring that information is protected from possible interception or eavesdropping by unauthorised persons.

Integrity: Integrity ensures that data is not altered or modified intentionally or accidentally. This helps to ensure the accuracy, reliability and durability of the data. The integrity of the data will only be guaranteed if it is protected from active eavesdropping which may modify the intercepted data. This protection can be achieved by implementing security mechanisms such as:

> rigorous access control;

> data encryption;

> protection against viruses, worms or Trojan horses;

Availability: The availability of services, systems and data is achieved by appropriate sizing

and redundancy of infrastructure components and by operational management of resources and services. Availability is measured in terms of the period of time during which the service offered is operational.
The potential volume of work that can be handled during the period of availability of a service determines the capacity of a resource (e.g. server or network). The availability of a resource is, moreover, inseparable from its accessibility.

> Ensuring that data and operational services are always accessible even in the event of a breakdown or attack.

> To ensure that these elements are accessible at the appropriate time by authorised persons.

> Maintain continuous accessibility without interruption or degradation;

3. Intranet security

The purpose of the intranet is to distribute internal documents, which may be confidential. It is designed according to the needs of the company. In addition, it allows a total reliability of the data, thanks to their content and their transmission. An intranet facilitates teamwork and the realisation of joint projects. To get the maximum benefit from an intranet, the system must be defined in terms of processes.

3.1 Intranet functions

The essential functions of an intranet are :

> Gathering information to perform a task.

> Facilitating the collaboration of dispersed individuals or units to achieve
of a common project.

> Reorganise tasks around a common process for several departments.

The intranet is an integrated management tool, allowing all employees to have a common working environment. The intranet provides its users with information and tools to help them in their daily work. It facilitates the integration of each employee into the overall company process.
It must allow a complete interaction between the components of the information system which is the Internet site, the extranet site and the integrated management of the company. All the actors of the company, whether they are employees, customers, suppliers or partners, interact on a single system which is articulated around the various processes of the firm.

3.2 Safety in relation to third parties

An adopted management of access rights to information must be put in place to guarantee the confidentiality and integrity of the information. Unlike the internet where security is poor, a well connected internet offers very good security.

- Management of access rights to data in a very secure manner.
- Fine-grained rights management: documents accessible only to certain people according to their function, status, etc.
- Impossible to penetrate the intranet from outside.
- Security is based on specific data metas. Indeed, document description standards seek above all to ensure the independence of the logical and physical representations of information. To define the security level of a document, other description standards are needed. They must specify a number of indicators:

> .2.1 For the administrator

The administration of the computer system ensures the management of the servers and the access rights to the data. It also manages users, user groups and their access rights

> .2.2 For the user

Within a computer system, users allow the possibility of description and management of data according to the access rights set by the system administrator.

4. The threats

They are determined adversaries capable of mounting an attack exploiting a vulnerability. [10]

These threats which are seen as potential sëcurity violations and which exist due to the vulnerabilities of the system can cause several effects such as:

Destruction of information and/or other resources ;

Corruption or modification of information; (see Annex: Figure A.7)

Theft, deletion or loss of information and/or other resources; (see Appendix Disclosure of information ;

Service interruption; (see Annex figure.1)

Threats can be classified as either ***accidental*** or ***intentional***.

4.1 Accidental threats

Accidental threats are those that exist without prëmëdication. It is difficult to cover the full range of accidental incident threats to corporate IT infrastructures. Certainly human error, system failures and computer programming errors or bugs in software can be mentioned.

Although major software manufacturers have committed to improving the security of the products they offer on the market, they still present a significant threat to the security of information networks due to insecure software.

4.2 Intentional threats

A person manages to get into the system, legitimately or not, and then access data or programs to which he or she is not supposed to have access, for example by using known and uncorrected flaws in the software

Intentional threats can be classified into two groups according to the damage each one causes. These two groups are ***passive*** and ***active threats***.

4.2.1 Passive Threats

Passive threats consist of eavesdropping without changing the data, neither the network operation nor the system state changes. It is very difficult to detect this type of threat because it is harmless compared to the normal functions of the system.

4.2.2 Active threats

Active threats or attacks consist of modifying data or messages, breaking into network equipment or disrupting the proper functioning of the network. An active attack can be executed without the ability to eavesdrop. Furthermore, there is usually no prevention possible for these attacks, although they are detectable (thus allowing an adequate response). (See Annex 2)

Figure 1 shows the different types of threats.

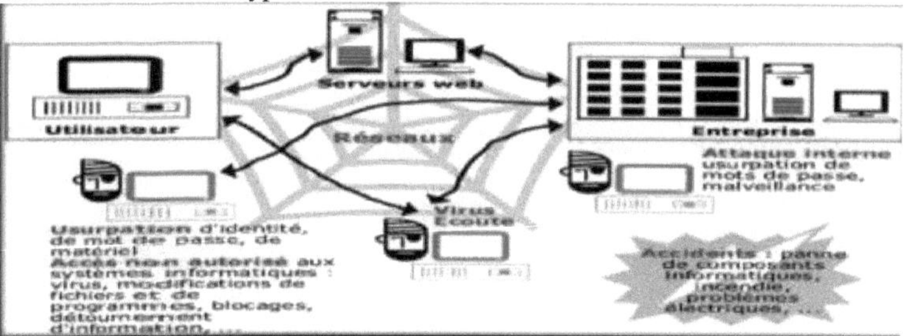

Figure 1: The different types of threat

5. The strategic security

The implementation of a security approach requires the realisation of the following points [4]:

Security management is specific to the organisational structure of the company and depends on its strategy. There are therefore as many security policies, procedures and tools as there are companies and security needs.

A simple, precise, understandable and applicable security policy.

The publication of the security policy.

Recording, monitoring and auditing procedures;

The expression, control and enforcement of safety clauses in different contracts;

Staff who are aware of and trained in security, and who have a high moral value;

The company's top management is responsible for assessing the risks, establishing the security policy and setting up the organisational structure to implement it. Risks and policy are assessed and updated.

A sëcuritë policy offers a graduated response to a specific proЬlёгe sëcurity, depending on the risk analysis made. It must express the balance between production and protection needs.

Centralized security management and some automation of security processes.

A level of trust determines the people, systems and tools involved.

6. Conclusion

In this chapter, we have presented the objectives of computer network security and we have examined the various factors that will affect the expansion and use of such networks.

In the next chapter, we will present a detailed study of the solution, as well as the installation and configuration steps of the tools used for data management. We will then proceed to the data implementation test phase.

CHAPTER 3

Study of the architecture

1. INTRODUCTION

In this chapter, we study and present an architecture adopted to the EAH firm's IT environment. We installed the GLPI asset management software and different servers such as the LMAP web server, the SAMBA and NFS file servers, the DHCP server and the FTP server.

2. Architecture adopted

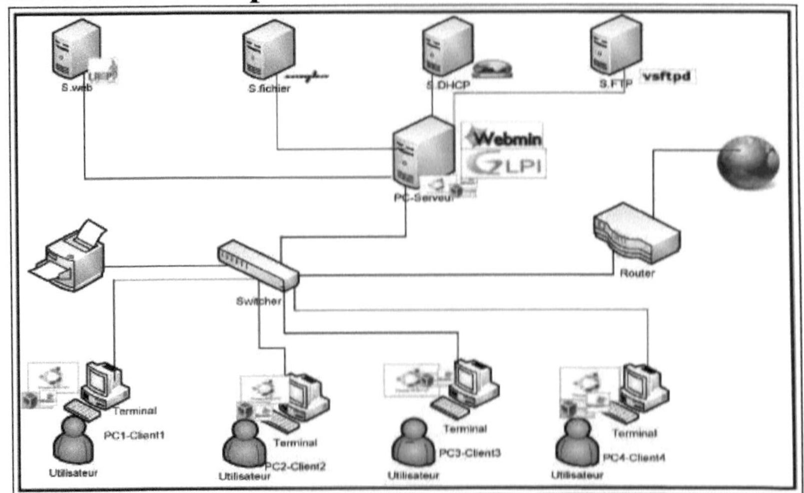

Figure 2: The network to be studied

Table 2: Characteristics of the network to be studied

Equipment	@IP	Mask
PC1-Client1	**192.168.1.100**	**255.255.255.0**
PC2-Client2	**192.168.1.101**	**255.255.255.0**
PC3-Client3	**192.168.1.103**	**255.255.255.0**
PC4-Client4	**192.168.1.104**	**255.255.255.0**
PC-Server	**192.168.1.110**	**255.255.255.0**
Router	**192.168.1.1**	**255.255.255.0**
Printer	**192.168.1.240**	**255.255.255.0**

3. Working Environment

During this project, which was carried out in the *EAH* cabinet, we used several hardware and software components:

3.1 Material environment

The material environment is constituted by :

> An SMC 7904BRA3 router with a 2 MO ADSL module.
> An ACTEL 24-port network switch
> Printer : Samsung 2571 n
> Five machines (one server and four clients), with the following characteristics

<u>*Server :*</u>

Brand : DELL

Processor: core 2 duo 2 GHz

RAM: 440 GB

Hard disk capacity:

DD1=160 GO

DD2 =160 GB

Customer:

Brand : DELL

Processor: core 2 duo 2 GHz

RAM: 240 GB

Hard disk capacity: 240 GB

3.2 Software environment

The main tools used in our work :

3.2.1 Operating system: "Ubuntu 10.04 LTS

Ubuntu ***"GNU/Linux"*** is a free and secure operating system based on Linux developed by communication. Based on the *Debian* Linux distribution and using the GNOME desktop, this distribution which combines stability and user-friendliness^. [12]

ubuntu10.04 LTS (Lucid Lynx) was released as a stable version on 29 April 2010. It is the twelfth version of the distribution, the third with long-term support *(LTS).*

LTS (Long Term Support) versions are Ubuntu versions that get pr^Пёдхё treatment in terms of stability and support. More attention is paid to the overall stabile of the ***"LTS"*** version as it gets technical support and sëcurity updates for a longer përiod of time.

3.2.2 Oracle VM virtualBox 3.2.8_OSE r64453

EAH uses accounting software^ that runs under Windows such as
such as "I compta", "Compta-100", "Ciel". For this, we will use virtualBox to satisfy the needs.

3.2.3 Web administration interface for Linux systems: "Webmin

a) Definition

Webmin is a web administration interface based on perl and allowing
No proprietary database that would prevent conventional intervention after installation. [13]

With a simple browser, you can, through webmin :

> Dëfinish user accounts ;

> Gëer rights to directories and files ;

> Sharing volumes ;

> Paramëtrer DNS;

> Use packages from servers such as: SAMBA, FTP, DHCP, FIREWALL...

Webmin supports the concepts of plug-ins and each service is managed through a webmin module. The list of services that can be administered by webmin is already very comprehensive as standard, but many special modules for managing less common services can be added.

The main features of the *webmin* are :

> Remote management of a server without using command lines;

> a multitude of modules that allow the configuration of most of the server's services;

> The control of the major software servers used by network administrators and their configuration locally or remotely works with any browser;

> Use of CGI scripts that directly update the system files;

b) Installation and Configuration

To install ***"Webmin"***, the configuration is done during the installation:

> Edit the file */etc/apt/sources.list'* and add the line :

deb http://download.webmin.com/download/repository sarge contrib
Sudo wget http://www.webmin.com/jcameron-key.asC
Sudo apt-key add jcameron-key.asc

> Now use the commands :

Sudo apt-get update then *sudo apt-get install webmin*

> To access "webmin" from a web browser, type the IP address of the server (192.168.1.110) followed by port 10000. An authentication interface with a secure connection is displayed asking for a username and password (see figure 3)

"https://@IP of the server:10000/"

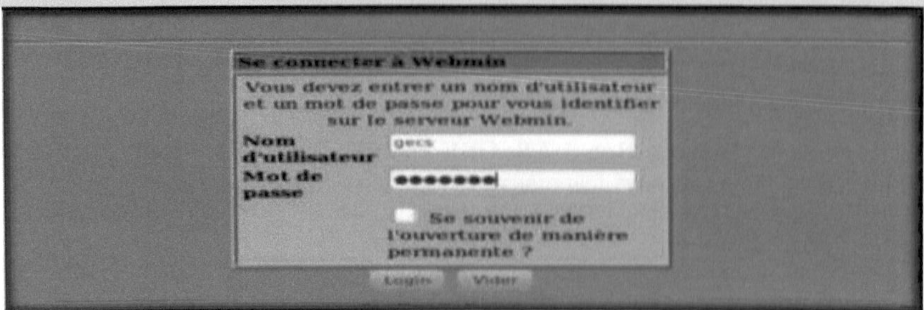

Figure 3: Connection to webmin

Figure 4 shows the homepage of the *webmin* with the different features.

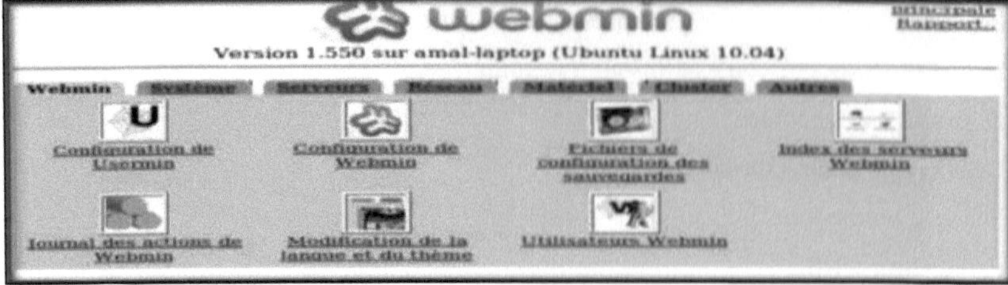

Figure 4: Webmin homepage

3.2.4 Open Source IT Asset Manager: "GLPI

GLPI,[20] An open-source IT asset management solution, is also a full web application for дёгег 1 set of IT asset management issues: from managing the inventory of hardware or software components of an IT asset to managing user support. [14]

a) Installation and Configuration [Case of EAH].

To install ***GLPI,*** you must first :

> Tëlëcharger version stable de *GLPI* ;

> Dëcompress using this command:

tar -xvzf glpi-0.78.5.tar.gz -C /var/www/

> Change the ownership of subdirectories and files included in the glpi directory by the Apache server using this command:

*Chown -R www-data : www-data /var/www/glp^*

> Install *GLPI* through the web browser;

> To access ***GLPI*** from a web browser, type :

http://localhost/glpi/ ***(e.g. http://192.168.1.110/glpi/).***

Figure 5 allows to test the configuration of the apache server and the existence of the MySql database manager, if all the test steps are correct, we can go to the next step.

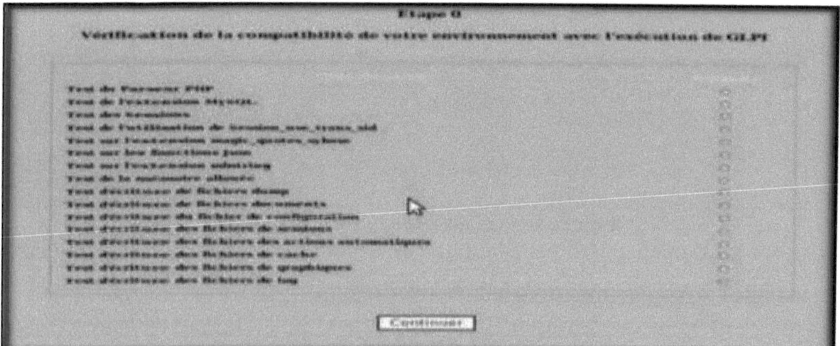

Figure 5: Checking the accounting of the installation environment

The figure 6 allows to configure the database used in GLPI (in our case the host is that of the address of the server machine (192.168.1.110).

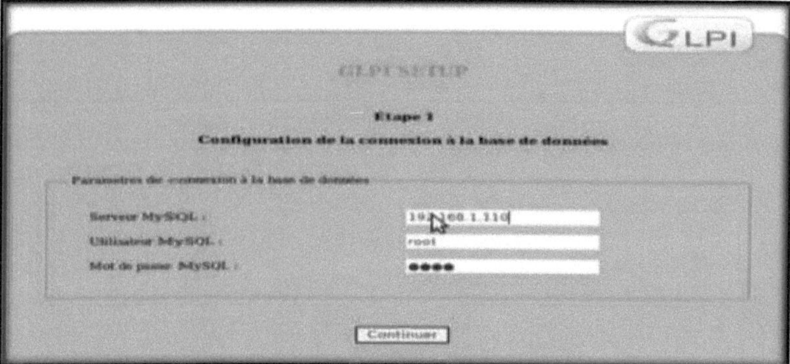

Figure 6: Database connection configuration

Figure 7 proposes to do the installation using an existing database or to create a new one. In our case, we have chosen to create a new database named "gecsparc".

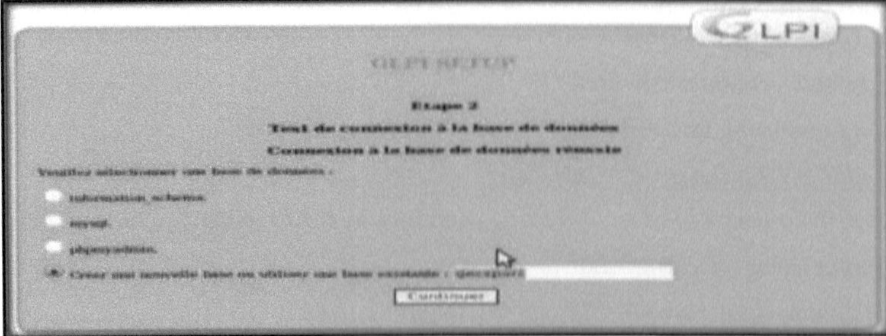

Figure 7: Database creation

Figure 8 allows the creation of a new base поттёе "gecs pare" нйНзё in GLPI, if the base is well тШаНзёс, we can go to the next step.

Figure 8: Initialization of the database

Figure 9 shows the default end of installation of all accounts (administrator, technician, normal, postonly), so can modify or delete these accounts.

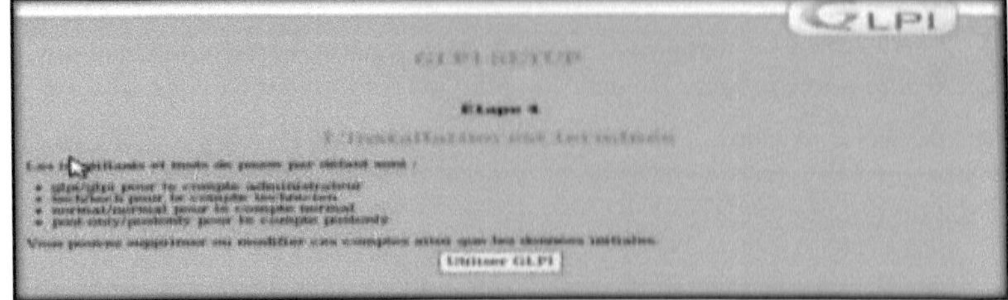

Figure 9: Installation

Figure 10 allows authentication to *GLPI*, using in our case as username "glpi" and password "glpi".

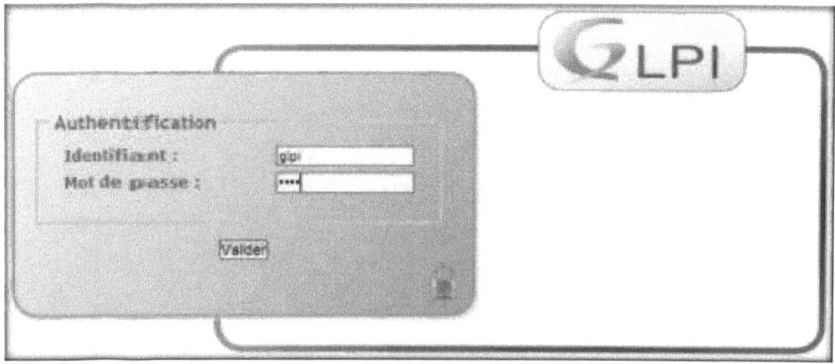

Figure 10: Authentication

Figure 11 shows the main page of the GLPI, it contains :

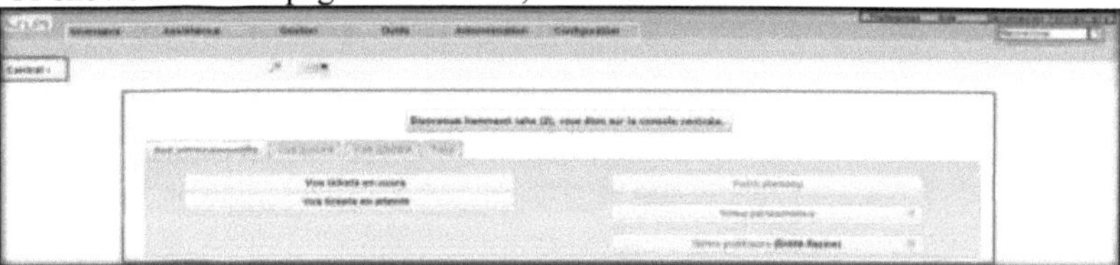

Figure 11: GLPI Main Interface

> **The User Preferences Menu:** The red boxed area at the top right of the page in Figure 12 represents the User *Preferences* menu. It allows you to edit the profile information of the user account with which you are connected. The *"help"* button redirects to the http://glpi-project.org/help-central website. The button *"disconnection"* makes it possible to close the session and redirects towards the interface of authentification seen before.

> **The search field:** the green boxed area just below the *"user preferences"* menu is a *search* field. It allows you to search by keyword for material elements, or virtual elements.

> **The main menu of GLPI :** the yellow framed zone corresponds to the menu of *GLPI.* Different entries are present in this menu, likewise each entry of the menu has submenus.

> **The banner of menue :** The pink zone represents the banner of the under menu. GLPI has a large number of links that redirect to other pages, this banner indicates the current section.

> **The main view :** The red zone of l'ëcran is the main one which allows on this view the menues of connections to *GLPI.* It visually presents all the information of which of the recipient or which concern him (example: ticket in progress, pending, notes, planning). It can display ëgalement this information.

***Example [EAH case]*:**

> Item configuration: ***Inventory -> Computers***

In the inventory section, the hardware and software inventory of EAH's computer equipment can be listed and consulted by ensuring different properties:

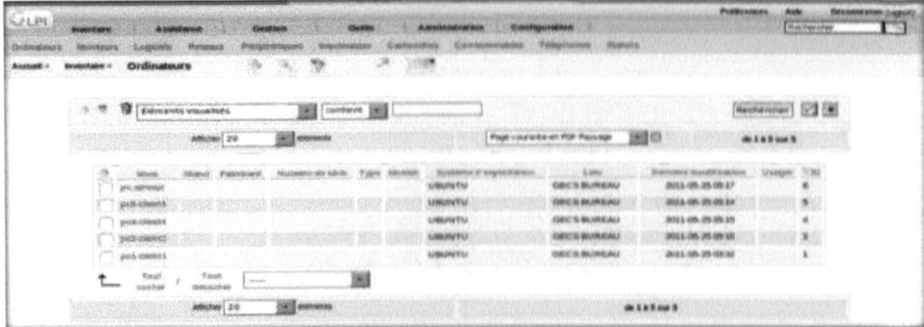

Figure 12: EAH machine inventory

> User Configuration*: **User Administration***

User management allows the creation, deletion or modification of users in the GLPI system.

To add a user, it is necessary to fill in the mandatory fields such as (Username, name, password, ca1ёдопе...). (See figure 13)

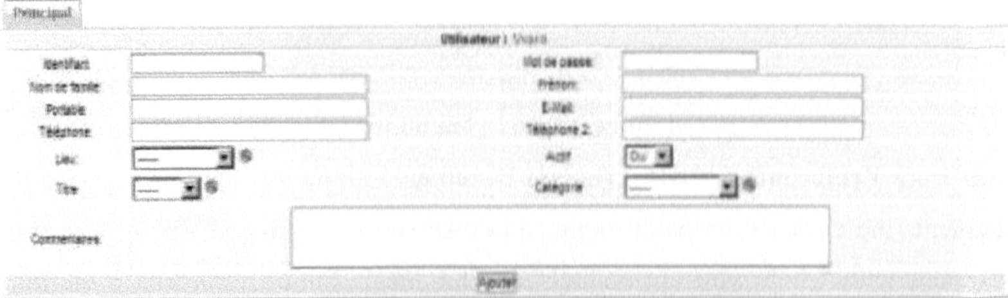

Figure 13: Adding a user from the EAH cabinet

Figure 14 shows the addition of a service category in the EAH cabinet.

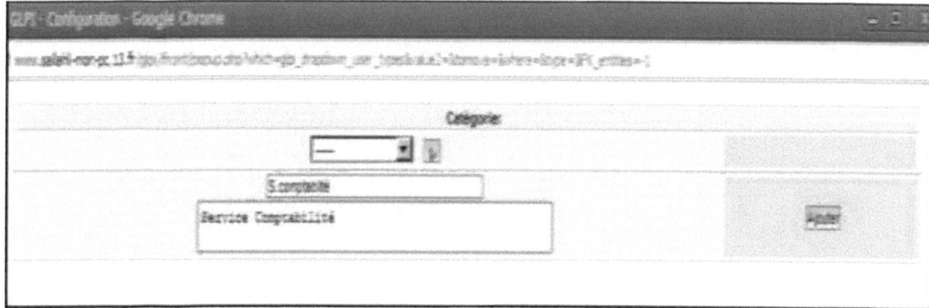

Figure 14: Adding the service category

> Case study : ***Support -> Tickets***

A ticket is a problem reported by a user and assigned to a registered technician.

The Tickets menu allows you to view the history of tickets and to keep track of current tickets.

To modify a ticket, simply click on the line that allows the administrator to add a technical support that will be sent from the requester, (see figure 15) which represents the interface for creating a new ticket with the following characteristics

We can also find several categories of hardware or software type to manage (add/modify/delete) such as OS version, PC brands, suppliers for hardware park also serial number, type, key for software park.

GLPI also allows to identify the titles of the personal or user.

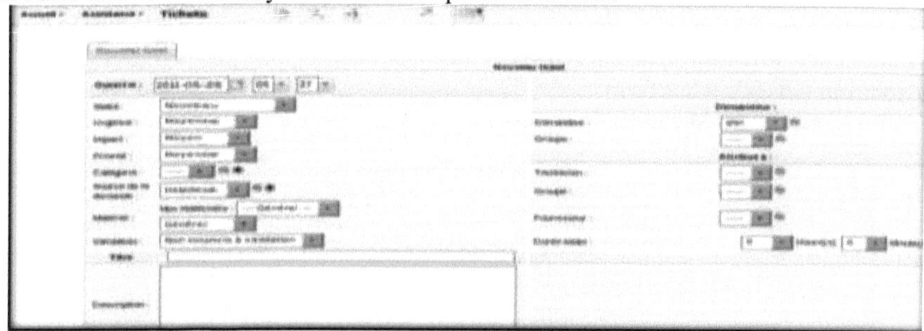

Figure 15: Adding a ticket from the EAH cabinet

Ticket - example

Status (new, assigned, planned, pending, closed resolved or unresolved)
Emergency (medium, high, very high, very low, low)
Priority: specifying the priority of the probes
C atëgory (s.commercial, s.accounting...)
S ource of the request (Help desk, Email, Other ..)

The interface in Figure 16 shows an example of a software failure from the EAH firm with the following characteristics:

Title of failure: **Software** (hardware or software)
Fault description: (See figure 16)

Figure 16: Software failure

Figure 17 shows the description of the software failure mentioned above.

Figure 17: Description of the fault solution

Title: Title of the problem, and in the field below, we can describe the details of this proЬlĕте.

Once all this information is complete you can send a message.

Figure 18 shows the breakdowns of all hardware and software in the firm's computer pool (EAH). Thus it allows to display the list of tickets according to certain criteria generally:

> Current tickets (assigned or scheduled).
> Non-firm tickets (all but firm).
> Firm tickets (resolved or not).
> All tickets.

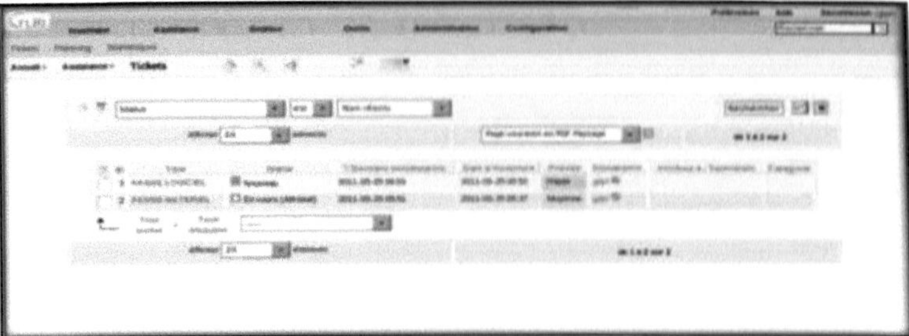

Figure 18: Trouble shooting

3.2.5 Software and tools used for enterprise security

Security tools and software will be used in our architecture and will be further developed in the following chapter: Firewall, Proxy, Backuppc, SSH.

4. Setting up the servers

In order to enrich the computer park of the EAH office, we have to install and configure different servers, namely: the web server (LMAP), the file server (SAMBA, NFS), the DHCP server and the FTP server.

4.1 Web server : Linux Apache Mysql PHP "LMAP

4.1.1 Definition

An *"http"* server or *"http daemon"* or *"HTTPd"* (HTTP daemon) or (less precisely) *web server*, is software serving requests respecting the client-server *HyperText Transfer Protocol (HTTP)*, which has ële dëveloppedë for the World Wide Web.[15] The HTTP daemon is a server that is used to serve requests from the Internet.

4.1.2 Installation

To install the "***web server", you need*** to install the necessary packages: Apache2, Mysql-server, Php5, **Php5-mysql** :

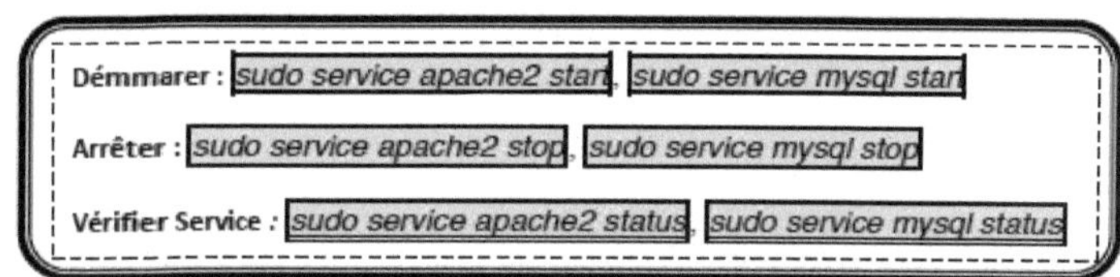

Sudo apt-get install apache2 mysql-server php5 php5-mysql
oëттarer: *sudo service apache2 start, sudo service mysql* **start**
Stop: *sudo service apache2* **stop|**, *sudo service mysql stop*
Vërify service: *sudo service apache2* **status|**, *sudo service mysql status*

4.2 File server: "SAMBA, NFS

4.2.1 SAMBA

a) Definition

"Samba is a *file server* for Linux (licensed under the GNU GPL). It is compatible with Microsoft Windows networks. It is called *SMB*. It allows [16]:

J The sharing of files and peripherals between lunix and Windows machines, i.e. between two different operating systems located in the same local network (home or company network).

b) Samba architecture

The server consists of two applications (called dëmons):

> **SMBD, the** core of the server, providing the following services:

J sharing resources.
J дёгег authentication (login and password).
> **NMBD**, allowing :
-*S* Show the services offered by samba.
J list all machines connected to the server.
J NETBIOS name resolution.

c) Installation and Configuration

To install the *"samba"* file server, the following packages must be installed directly required:
*Sudo apt-get install system-config-samb^*
Now we will see through *"webmin"* the configuration of the *"samba"* file server as shown in figure 19.

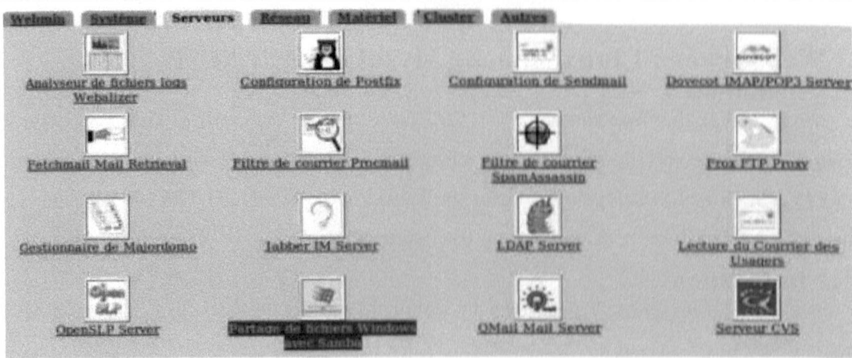

Figure 19: Samba server home

Figure 20 shows the home page of the *"Samba"* file server by first clicking on the *"Servers"* tab. Then, select the *"Windows File Sharing with Samba"* icon to create a new file or device share (See figure 21)

Informations sur le partage
Nom de partage
Répertoires personnels partagés
Répertoire à partager
Disponible ? Oui Non
Affichable ? Oui Non
Commentaire sur le partage
Autres options des partages
Sécurité et contrôle d'accès
Droits d'accès des fichiers
Nommage des fichiers
Options diverses

Figure 20: Creating a new Share file

Figure 21 shows the list of shares available on the server, let's quote the example of "Document Accounting" under the folder /var/doc-accounting, the right as shown in the figure is read-only for all users, that means that this share is only for model documents.

Sélectionner tout. | Inverser la sélection. | Créer un nouveau fichier partagé | Créer une nouvelle imprimante partagée | Créer une nouvelle copie | Voir toutes les connexions

	Nom de partage	Chemin	Sécurité
☐	printers	*Toutes les imprimantes partagées*	Utilisable par tous les utilisateurs connus
☐	print$	/var/lib/samba/printers	Lecture seule pour tous les utilisateurs connus
☐	DocumentComptable	/var/doc-comptable	Lecture seule pour tous les utilisateurs connus

Figure 21: File sharing of the EAH cabinet

Figure 22 represents different files or përiphëriques shared through the "EAH" cabinet.

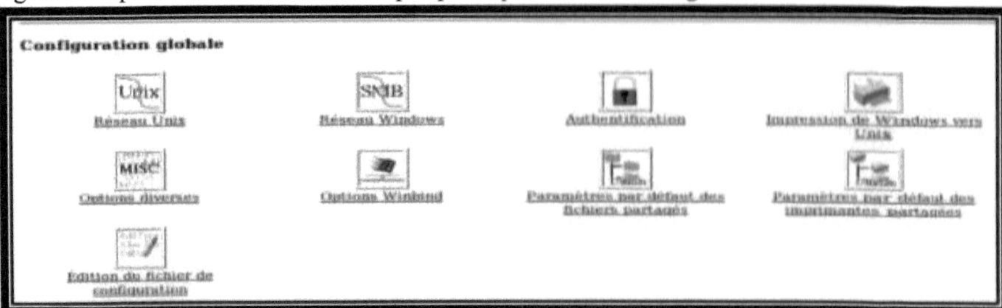

Figure 22: Global configuration

In the *"Global Configuration"* section, click on *"Windows Network"*. After that, we have to check if the *workgroup* corresponds to the one of the computers under Windows. Also, we must check the *name of the server* that will be displayed on the network machines (see figure 23).

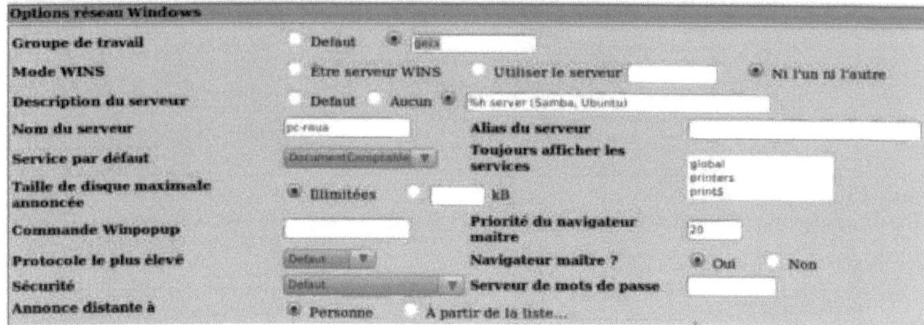

Figure 23: Checking the Global Configuration

In the *"Global configuration"* section, click on *"Edit configuration file"* to study the configuration files:

vi /etc/samba/smb.conf

For User and Group Management :

First we need to create the users who will have access to the file shares.

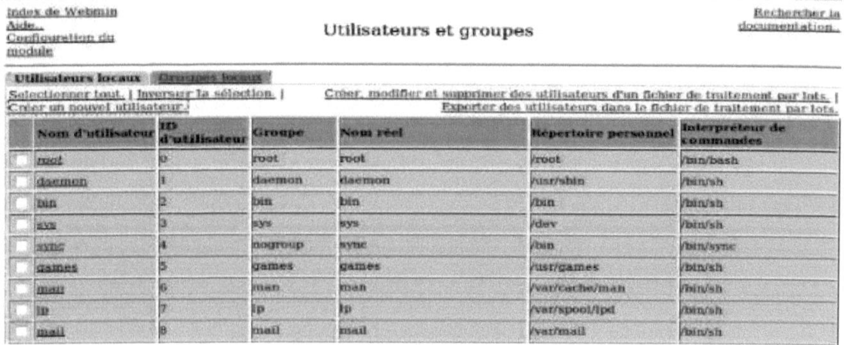

	Nom d'utilisateur	ID d'utilisateur	Groupe	Nom réel	Répertoire personnel	Interpréteur de commandes
	root	0	root	root	/root	/bin/bash
	daemon	1	daemon	daemon	/usr/sbin	/bin/sh
	bin	2	bin	bin	/bin	/bin/sh
	sys	3	sys	sys	/dev	/bin/sh
	sync	4	nogroup	sync	/bin	/bin/sync
	games	5	games	games	/usr/games	/bin/sh
	man	6	man	man	/var/cache/man	/bin/sh
	lp	7	lp	lp	/var/spool/lpd	/bin/sh
	mail	8	mail	mail	/var/mail	/bin/sh

Figure 24: Users and Groups

In the *"System"* tab of the menu, select the "Users and Groups" icon, in the *"Local Users"* tab, click on "*Create*" and fill in the "User Name" and "Password" fields, then click on *"Create"* (see figure 25)

Index de Webmin
Index du module
Aide..

Création d'un utilisateur

Détails de l'utilisateur
Nom d'utilisateur
ID d'utilisateur — Automatique / Calculé / 1005
Nom réel
Répertoire personnel — Automatique / Directory
Interpréteur de commandes
Mot de passe — Aucun mot de passe requis / Aucune connexion autorisée / Mot de passe normal / Mot de passe pré-encrypté
Nom de connexion temporairement désactivé

Figure 25: User creation

In the *"Local groups"* tab, click on *"Create a new group"*, and fill in the "Group name" and "Members" fields (see figure 26)

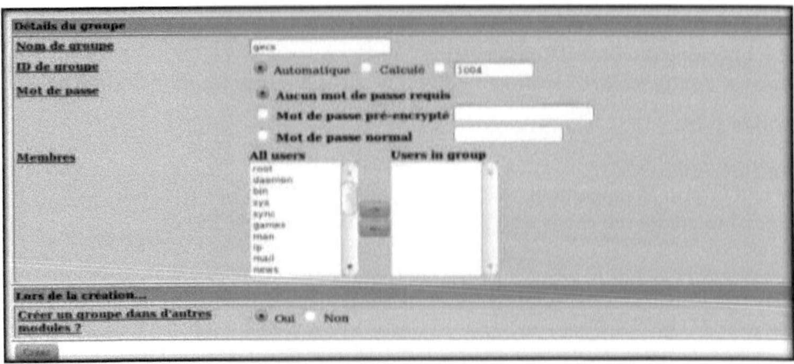

Figure 26: Creation of a group

In the *"Samba Users"* section, click on *"Convert Unix users to Samba users"* and on *"Convert users"* (see figure 27)

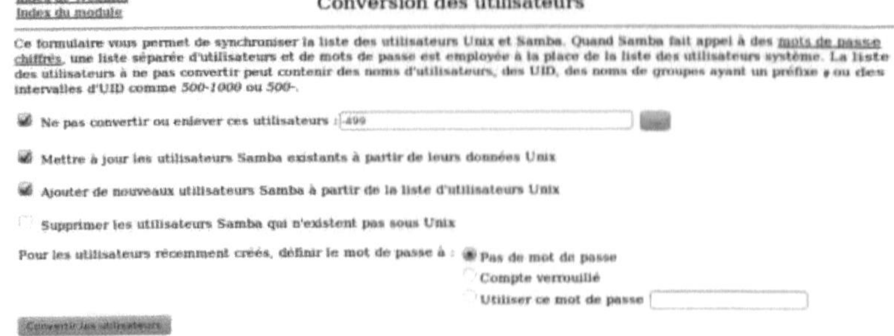

Figure 27: Conversion of users

After all changes, the "*Samba*" server must be *"restarted"* by clicking on *restart Samba servers* or *stopped* through *"webmin"*. (See figure 28)

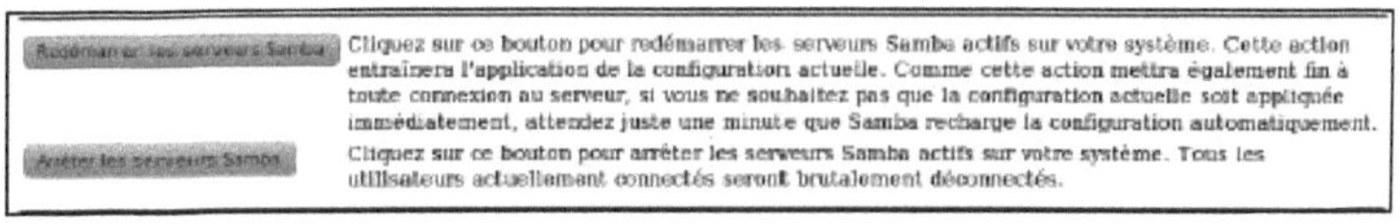

Redémarrer les serveurs Samba | Cliquez sur ce bouton pour redémarrer les serveurs Samba actifs sur votre système. Cette action entraînera l'application de la configuration actuelle. Comme cette action mettra également fin à toute connexion au serveur, si vous ne souhaitez pas que la configuration actuelle soit appliquée immédiatement, attendez juste une minute que Samba recharge la configuration automatiquement.

Arrêter les serveurs Samba | Cliquez sur ce bouton pour arrêter les serveurs Samba actifs sur votre système. Tous les utilisateurs actuellement connectés seront brutalement déconnectés.

Reboot Samba Servers I Click this button to reboot the active Samba servers on your system. This action will cause the current configuration to be applied. As this action will also terminate all connections to the server, if you do not want the current configuration to be applied immediately, just wait a minute for Samba to reload the configuration automatically Click this button to stop the active Samba servers on your system. All currently connected users will be abruptly disconnected.

Figure 28: Change of state

Or through the command lines:

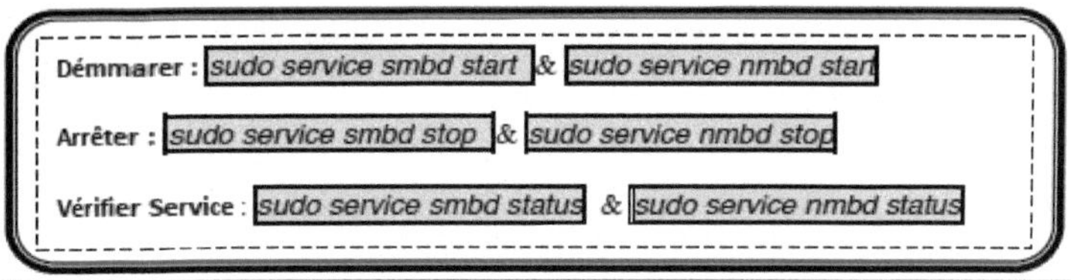

Démmarer : sudo service smbd start & sudo service nmbd start

Arrêter : sudo service smbd stop & sudo service nmbd stop

Vérifier Service : sudo service smbd status & sudo service nmbd status

Dëmmarer: *sudo service smbd start* & *sudo service nmbd start* **Stop:** *sudo service smbd stop* | & *sudo service nmbd stop* **Vërify Service:** *sudo service smbd status* & *sudo service nmbd status*

4.2.2 NFS

a) Definition

"**NFS** stands for Network *File* System.

It allows [18] :

J File sharing (folder sharing) between 2pc of the same system Operation.

J Sharing data with multiple users (user management).

b) Installation and Configuration

To install the *"NFS"* server, type the command :

Sudo apt-get install nfs portmap nfs-kernel-server-common

With :

nfs-kernel-server : NFS Linux Server ;

nfs-Common: NFS common programs ;

Portmap : The RPC port mapper

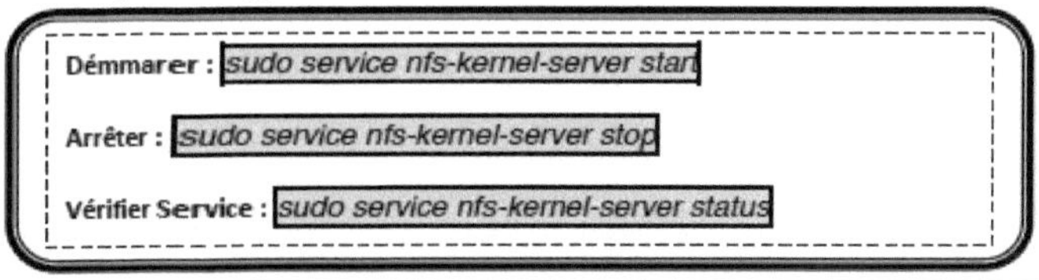

Démmarer : sudo service nfs-kernel-server start

Arrêter : sudo service nfs-kernel-server stop

Vérifier Service : sudo service nfs-kernel-server status

Start: *sudo service nfs-kernel-server start* **Stop:** *sudo service nfs-kernel-server stop*

verifier Service: *sudo service nfs-kernel-server status*

4.3 DHCP" server

4.3.1 Definition

"DHCP" stands *for Dynamic Host Configuration Protocol and* is a TCP/IP network service

that allows client computers to be configured automatically from a server.

A network protocol whose role is to provide automatic configuration of a station's IP settings, including automatically assigning an IP address and subnet mask. "*DHCP* can also configure the default gateway address, DNS and NBNS name servers (known as WINS servers for Microsoft networks)[21].

The most common configuration elements provided by a DHCP server to a DHCP client include: (IP address and network mask, DNS, WINS)

However, a *"DHCP"* server can also provide configuration items such as: Host Name, Domain Name, Default Gateway, Time Server, Print Server.

The advantage of using a *DHCP* server is the automatic configuration of the IP addresses. The advantage of using a DHCP server is the automatic configuration of the IP addresses of the network, e.g. a change of address of a DNS server only needs to be changed on the DHCP server. All network clients will be reconfigured the next time they connect to the DHCP server. Another advantage is the easier integration of new computers into the network, since there is no need to check the available IP addresses.

4.3.2 Installation and Configuration

To install the ***"DHCP"*** server through the command line:

```
Sudo apt-get install dhcp3-servei
```

Through *webmin,* we can configure ***"DHCP"*** in the *"Server"* tab of the menu, we choose the *"DHCP server"* icon, we create a network by defining its parameters (SR description, @IP, SR mask, @ range...)

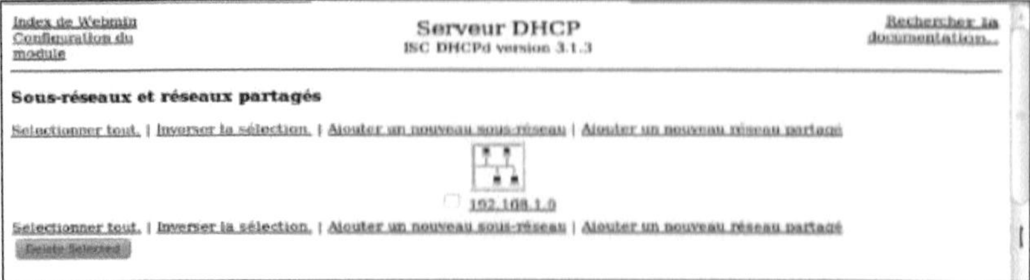

Figure 29 : Creation of a sub-network

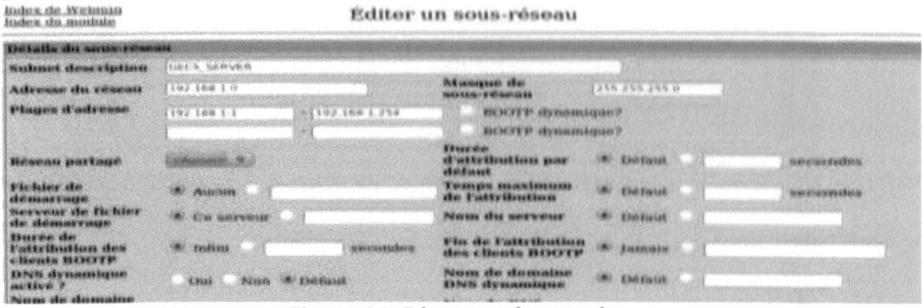

Figure 30: Editing a subnetwork

we add all the machines of the "EAH" cabinet's computer wall, ensuring the various parameters (description, machine name, Ethernet type hardware, IP...)

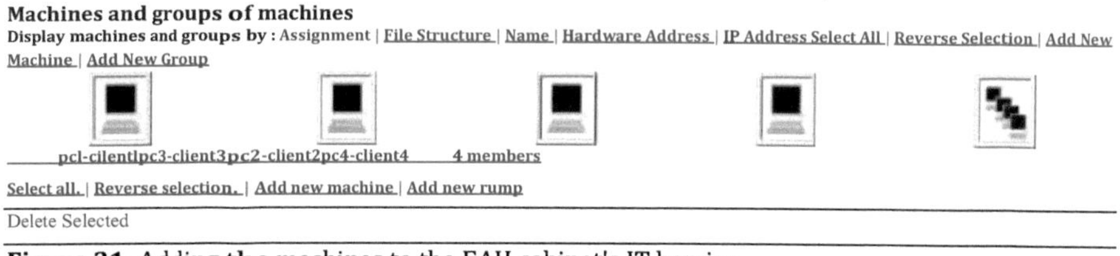

Machines and groups of machines

Display machines and groups by : Assignment | File Structure | Name | Hardware Address | IP Address Select All | Reverse Selection | Add New Machine | Add New Group

pcl-cilentlpc3-client3pc2-client2pc4-client4 4 members

Select all. | Reverse selection. | Add new machine | Add new rump

Delete Selected

Figure 31: Adding the machines to the EAH cabinet's IT barrier

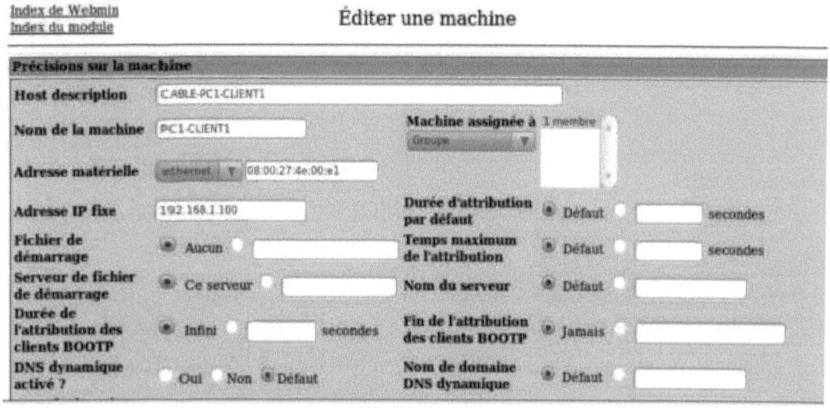

Index de Webmin
Index du module

Éditer une machine

Précisions sur la machine

Host description	CABLE-PC1-CLIENT1		
Nom de la machine	PC1-CLIENT1	**Machine assignée à**	1 membre Groupe
Adresse matérielle	ethernet 08:00:27:4e:00:e1		
Adresse IP fixe	192.168.1.100	**Durée d'attribution par défaut**	Défaut secondes
Fichier de démarrage	Aucun	**Temps maximum de l'attribution**	Défaut secondes
Serveur de fichier de démarrage	Ce serveur	**Nom du serveur**	Défaut
Durée de l'attribution des clients BOOTP	Infini secondes	**Fin de l'attribution des clients BOOTP**	Jamais
DNS dynamique activé ?	Oui Non Défaut	**Nom de domaine DNS dynamique**	Défaut

Figure 32: Editing

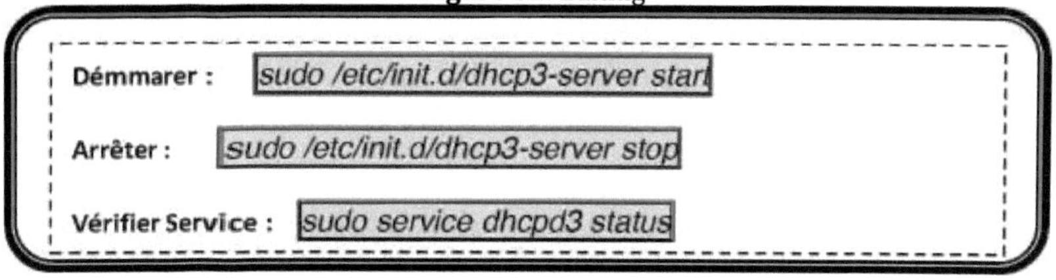

Démmarer : `sudo /etc/init.d/dhcp3-server start`

Arrêter : `sudo /etc/init.d/dhcp3-server stop`

Vérifier Service : `sudo service dhcpd3 status`

4.4 File transfer server: "FTP

4.4.1 Definition

The File Transfer **Protocol** (***FTP***) is *a protocol for transferring data* over a TCP/IP network. The implementation of *FTP* dates back to 1971, when a mechanism for transferring files between machines at MIT (Massachusetts Institute of Technology) was developed. It is the optimal protocol for the exchange of files between users. [24]

The objectives of the *"FTP"* protocol are to [11] :

allow file sharing between remote machines;

allow indëpendence on the file systems of client machines and

server ;

enable the efficient transfer of data;

4.4.2 Installation and Configuration

To install the ***"FTP"*** server, type :

Sudo apt-get install vsftpd
FTP server configuration is done by ëditing the file /etc/vsftpd.conf

To ëdit this file, do:

gksu gedit/etc/vsftpd.conf

> Prohibit "anonymous FTP", because in this case remote clients can access the FTP server *without authentication.*
In this case, the line :
anonymous enable=YES by: *anonymous enable=NO*
> Allow local users to log in by typing :
local enable=YES
This will allow users dëclarës in Ubuntu to use the same login/password to access the FTP server.

> Allow users to tëlë download files by typing :
write enable=YES
>Compile mode (Test/verify) *Chkconfig vsftpd on*

Demmarer: *sudo /etc/init.d/vsftpd start*
Stop: *Sudo /etc/init.d/vsftpd stop*
check service: *sudo /etc/init.d/vsftpd status*

The *"FTP"* ***client*** is the program that allows you to retrieve and/or send files to an *FTP* server with a graphical interface such as FILEZILA, a very good *FTP* client with a wxGTK interface, it is very versatile and easy to use. Its main drawback is that it saves all its passwords in clear text in the home directory.

5. Conclusion

In this chapter, we have adopted an architecture for the EAH firm that allows it to manage the data of its computer park. We have introduced different servers, namely the LMAP web server, the SAMBA and NFS file servers, the FTP server and the DHCP server.

In the next chapter, we will secure this architecture with strategies and tools to protect against various attacks and meet the needs of the EAH firm.

CHAPTER 4

Application of security to our architecture

1. INTRODUCTION

In this chapter, we will secure the EAH firm's IT environment based on well-studied strategies and tools to protect against attacks and meet our needs.

2. Securing the architecture

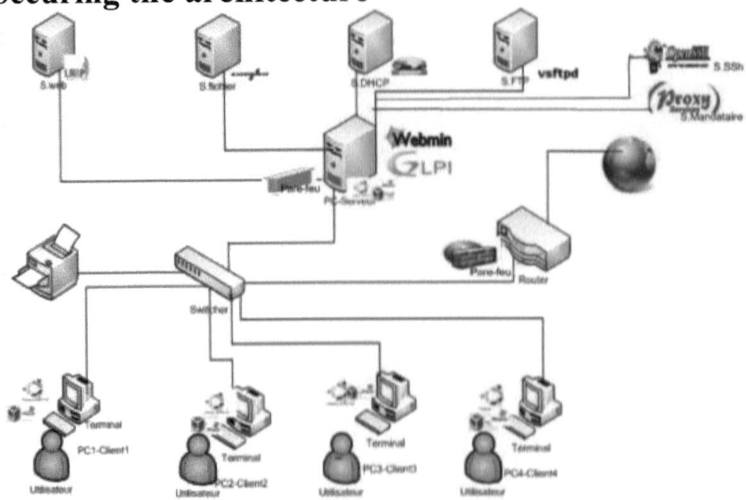

Figure 33: Architecture adopted

3. The different tools used for security

There are several solutions developed to secure computer network infrastructures against the numerous hacking and intrusion attempts. In the following, we mention some tools:

3.1 Proxy server: PROXY (SQUID3)

3.1.1 Definition

A *"proxy"*, or *"proxy server"*, is a computer server whose role is to relay requests between a client and a server. On the other hand, *"the proxy"* is originally a machine acting as an intermediary between computers on a local network and the Internet. So, most of the time the proxy server is used for the web, it is then an HTTP proxy. However, there may be proxy servers for each application protocol (FTP, ...).

Proxies are used to provide the following functions, among others:

J **Cache memory**: the ability to store (cache) the most frequently visited pages by users on the local network in order to provide faster access and reduce network load.

J **Filtering and anonymity ("filter requests, guarantee anonymity")**: This allows connections to be monitored via the creation of activity logs by systematically recording users' requests when they connect to the Internet. Thus it is possible to filter connections to the Internet by analysing client requests on the one hand, and server responses on the other. When the filtering is carried out by comparing the client's request with a list of authorised requests, it is called a white list, when it is a list of prohibited sites it is called a black list. Finally, the analysis of server responses according to a list of criteria (keywords, etc.) is called content filtering.

Authentication: Since the proxy is the indispensable intermediary between the users of the internal network and the external resources, it is sometimes possible to use it to authenticate users, i.e. to ask them to identify themselves using a user name and a password.

-*S* **Reverse Proxy:** Also known as a reverse relay, this is a proxy server that does not allow users to access the Internet, but allows Internet users to access certain internal servers indirectly.

3.1.2 Installation and Configuration

The installation of the ***"proxy server"*** is done by these steps:

Install ***Squid 3***: /usr/bin/apt-get install squid3

Through webmin, we type in the "server" tab of the menu, we have to choose the "server proxy Squid" icon. We click on "Ports and network" and note that the port is that "Squid" will use (by default port: 3128). (See figure 34)

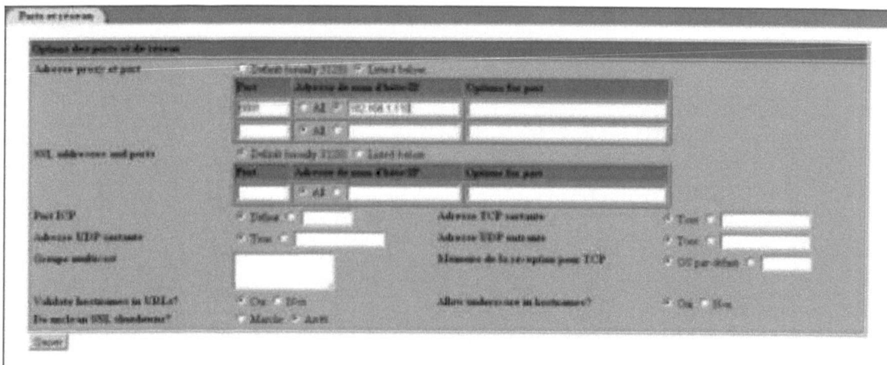

Figure 34: Squid ports and IP address

The address of the host name represents the IP of the machine on which we have installed the proxy server and the connection port, the choice of the port is arbitrary (depending on availability), in our case we have chosen 8081.

Once Squid is installed, it is necessary to configure it as a proxy on the client machines:

For Ubuntu client machines, simply configure the proxy server as shown in Figure 35

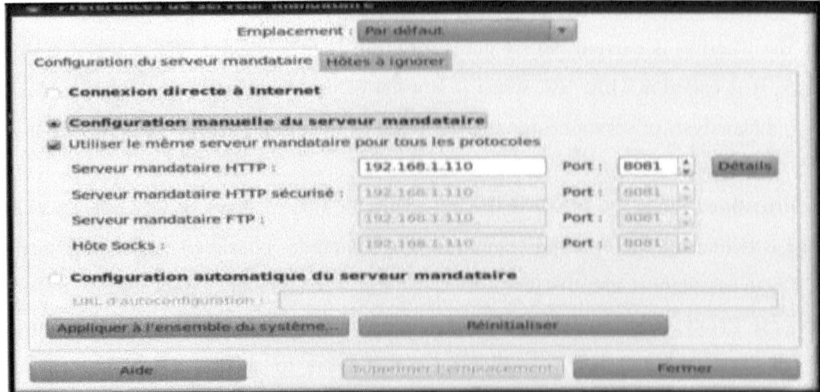

Figure 35: Proxy usage in unix machines

For Windows client machines, this is done by the following steps (see Figures 36, 37, 38):

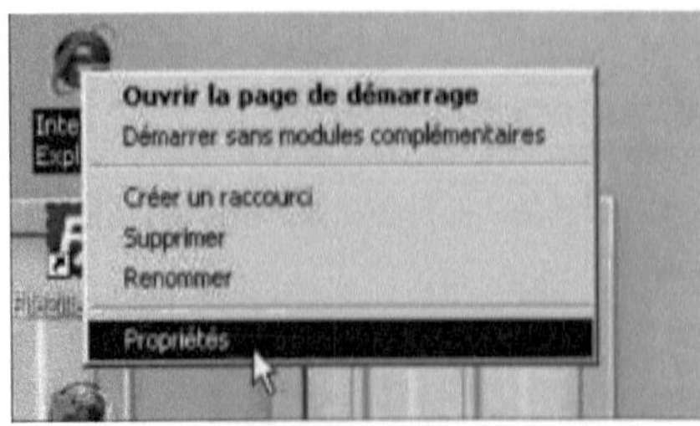

Figure 36: PROXY WINDOWS STEP 1

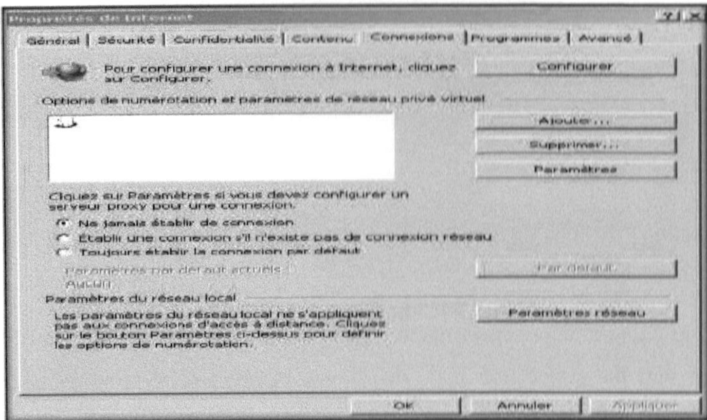

Figure 37: Proxy windows Step 2

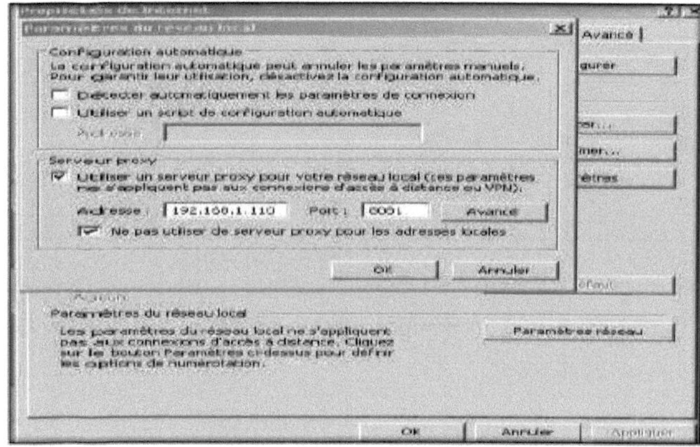

Figure 38: Proxy windowsStep3

These 3 steps allow the configuration of the proxy in windows machines.

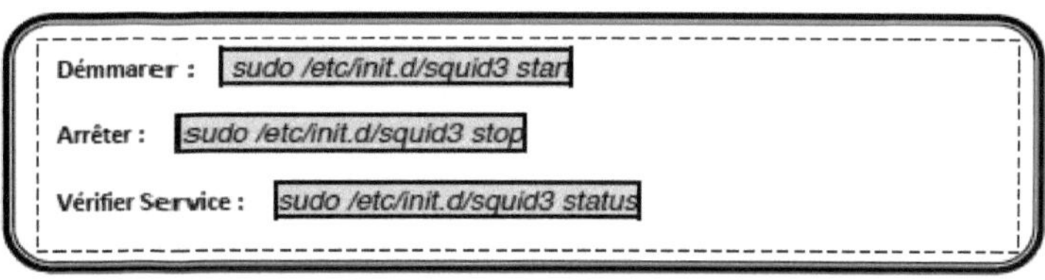

3.2 Netfilter/IPtables

3.2.1 Firewall

"***A firewall***" (or *firewall, security barrier* or *firewall*) is an essential enterprise network computing device comprising hardware (router(s), server(s)) and software (a paraтëTrer or to be dëveloped), which filters the flow of information between an internal network and an external network.

Its purpose is to protect the internal network from unauthorised access and actions from the outside, by controlling incoming traffic. The firewall can also control outgoing traffic.

There are three types of firewall techniques:

Jthe **packet filtering technique or firewalls** that aim to

block all traffic except that which is sëlectnë ;

Jthe **technique of proxy servers** which aim to achieve the

network connections as a proxy ;

J the **technique of gateways** that provide security systems to establish

TCP/IP connections between outside and inside or for certain services such as FTP and Telnet ;

Figure 39: Shema of a network structure protected by a firewall

The Linux kernel includes the Netfilter subsystem, which is used to manipulate or decide the fate of network traffic directed to or through your server. All modern Linux firewall solutions use this system for packet filtering.

When a packet arrives at the server, it will be passed to the Netfilter subsystem for acceptance, handling, or rejection based on the rules provided to it from userspace via IPtables. Done, the power and flexibility of Netfilter is implemented through the IPtables interface.

It allows :

- *S* Perform packet filtering, primarily to provide

Firewall.

- *S* Perform NAT (Network Address Translation) operations, These functions

are particularly useful when you want to communicate all or part of a private network, set up with private IP addresses, with the Internet.

- *S* Carry out marking operations on packages, to apply a

special treatment.

3.2.2 Netfilter Tables

Netfilter uses three tables which are :

> **SLhe FILTER table** which contains all the rules that allow to filter the

packages. It contains 3 strings:

> **The INPUT string:** this string will decide the fate of packets entering the host.

> **The OUTPUT chain:** concerns packets sent by the local host.

> **The FORWARD chain:** filters packets that traverse the host according to the implemented routes.

JThe **NAT (*Network Address Translation*) table:** This table allows non

It is not only possible to do strict address translation, but also port translation, it allows to protect the machines of the private network from direct attacks since they are, in fact, not accessible from the outside. On the other hand, one can restrict access to the network to certain machines and deprive others.

> **The PREROUTING string**: used to translate the destination address.

> **The POSTROUTING string:** allows the source address to be translated.

> **The OUTPUT string:** allows the destination of packets to be changed locally by the bridge itself.

3.2.3 Installation and Configuration

To install ***"firewall"***, we will use *"webmin"*, first in the *"Network"* tab of the menu, we choose the *"Linux Firewall"* icon (see figure 40)

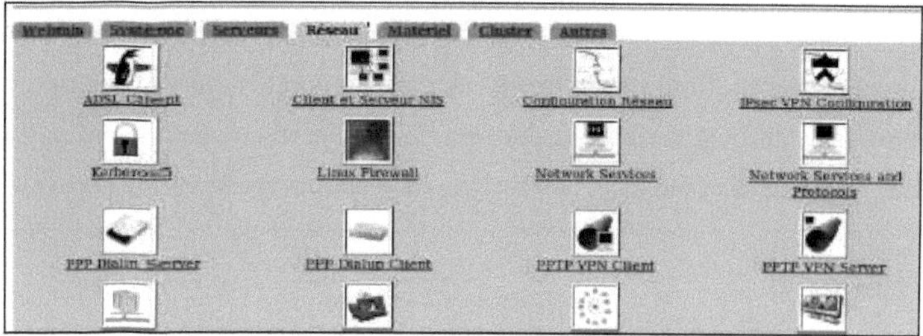

Figure 40: Firewall Home

After the installation one can add ***an "INPUT string rule"*** which allows to decide the fate of packets by entering different configuration parameters (String Part, Rule Comment, @IP type Equal, ICMP packet type, @Ethernet type Equal). Figure (....) shows an example

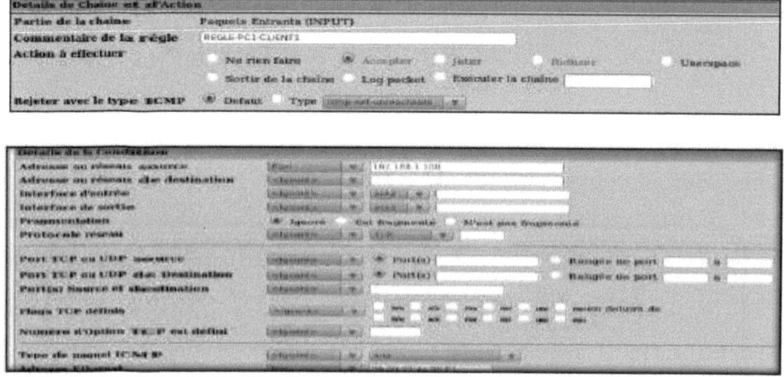

Figure 41: Studying a region

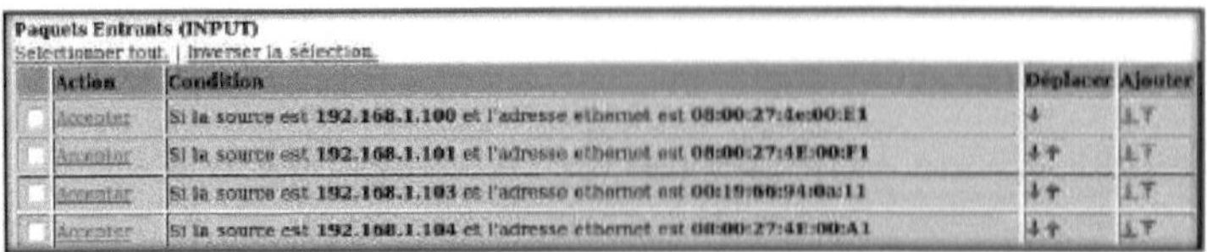

Figure 42: The list of regions

3.2.4 Firewall configuration at the router

*"**A router is**"* an interconnection device that routes packets between two or more networks. Its role is to forward packets from one network interface to another, according to a set of rules forming the routing table. In the case of the firm *"EAH"*, at the router level, as shown in figure 39, all accesses have been blocked except for the one coming from the server IP (192.168.1.110)

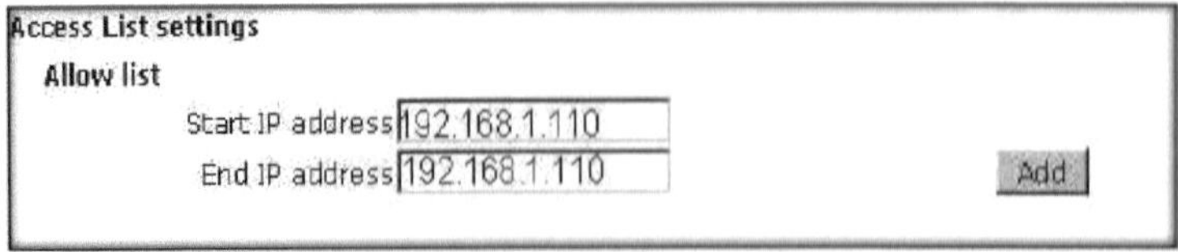

Figure 43: Firewall and router

3.3 "SSH" server

3.3.1 Definition

The ***"SSH"*** (Secure Shell) protocol allows you to connect to a remote machine with a secure link. Data is encrypted between machines. It allows commands to be exëcuted on a remote server. [21]

3.3.2 Installation

sudo apt-get install ss or *sudo apt-get openssh-server*

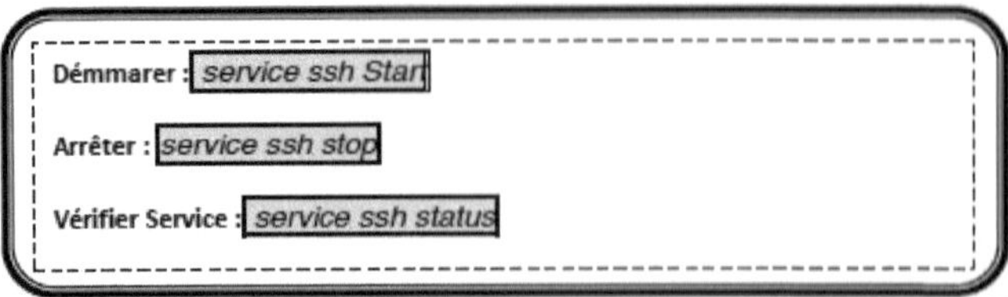

4. Conclusion

In this chapter, we have sëcurisë the computer park of EAH. we have implemented and configured the following tools: SQUID3 as a proxy server, Netfilter/Iptables as a firewall, SSH to secure remote access.

GENERAL CONCLUSION AND PERSPECTIVES

In this final year project, we have adopted a security architecture for the EAH's IT environment, against the various factors that threaten private data and information. To do so, we have studied computer security and we have proposed different tools and strategies to meet EAH's needs, such as GLPI, an Open Source IT Asset Management tool. It allows the management of the inventory of the hardware and software components of the park to the management and assistance of the users. We installed the LMAP Web server to ensure client-server communication, the SAMBA and NFS file servers for sharing files and peripherals between machines, the DHCP server for configuring machine IP addresses and the FTP server for data transfer.

To ensure security, we have introduced a firewall to protect the internal network against unauthorised access and actions from the outside, by controlling traffic, a proxy server to relay requests between a client and a server and an SSH server for remote connection with a secure link.

After this rich experience, we can prove our pride in working in this advanced field and mastering the strategies of securing computer parks in order to protect them from different possible attacks.

As a perspective of this work, we can extend the security of the park by integrating the SNORT server which detects network attacks based on a very up-to-date signature database (IDS), the LDAP server allowing to manage directories and to secure the access to the information stored in the database, thus, it proposes encryption (SSL) and authentication (SASL) mechanisms and the BACKUPPC data backup server to ensure the protection of the park's data against loss.

BIBLIOGRAPHIES

REFERENCES

- [1] Daniel Guinier, "Sëcuritë et qualité des systëmes d'information" Approach systëmique, Masson, 1992
- [2] Belgiu, "AWT (Agence Wallonne des Telecommunications)
- [3] Jean-Frangois Pilou, "Tout sur la securite informatique", Dunod 2009.
- [4] Didier Godart, "Securite informatique : risques, strategies et solutions" 2002
- [5] Gilles Chamillard, Sebastien Bobillier "Ubuntu - System and network services administration

WEBOGRAPHIES

- [6] http://www.fr.wikipedia.org/wiki/Informatique
- [7] http://wikipedia.org/wiki/Reseau_informatique
- [8]http://www.wikipedia.org/.../Technologies_de_l information_and_communicati on
- [9] http://www. en.wikipedia.org/.../Information_System_Security
- [10] http://www. en.wikipedia.org/.../ Computer threats
- [11] http://www.ftp.com
- [12] http://forum.ubuntu-fr.org/
- [13] http://www.webmin.com
- [14] http://www.ANSI.tn/
- [15] http://www.serveurweb .org/
- [16] http://www.samba.org/
- [17] http://www.nfs-server.org/
- [18] http://www.dhcp3-server.org/
- [19] http://www.netfilter.org/
- [20] http:// www.glpi-project.org
- [21] http://www.ssh.fr/
- [22] http://www.fr.wikipedia.org/wiki/Proxy
- [23] http://www.free-proxy.org/

[24] http://www.fr.wikipedia.org/wiki/File_Transfer_Protocol

Annexes

Annex 1: "GLPI

Figure A.1 allows the paraтё^aдe of the "GLPI" by sëlection the language that can be chosen.

Figure A.1: Installing language

Figure A.2 shows the terms of the *GLPI* licence that we accept.

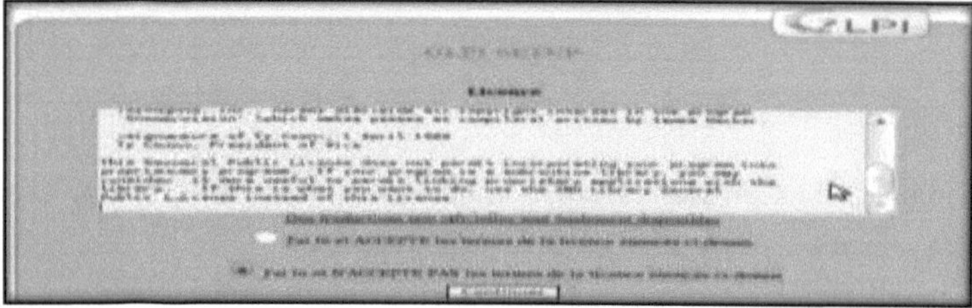

Figure A.2: The terms of the licence

Figure A.3 shows the types of installation either for a new compete installation , or to update your version from an earlier version.

Figure A.3: Types of installation

Annex 2: "THE MANACES

> ***The interruption***

This attack is aimed at availability, and occurs when one of the system's components is disabled (e.g. a virus attacking the hardware) or blocked (e.g. bandwidth saturation), i.e. the flow of information cannot go from the source to the destination.

Figure A.4: The interruption

> *Capture*
This attack targets confidentiality, and occurs when the attacker gains access to the transferred information. During the exchange, the attacker obtains illegitimate access to the information flow, for example by eavesdropping on the network (sniffing), by modifying the routing (RIP, OSPF, etc.) or addressing (ARP or DNS)

Figure A.5: Capture

> *Injection*
This attack is aimed at integrity, and occurs when the attacker injects his own information into the legitimate stream without it being noticed by the destination.

Figure A .6: Injection

> *The modification*

This attack targets confidentiality and integrity, and occurs when the attacker intercepts the stream emitted by the source and modifies it before sending it back to the destination in the role of the man in the middle.

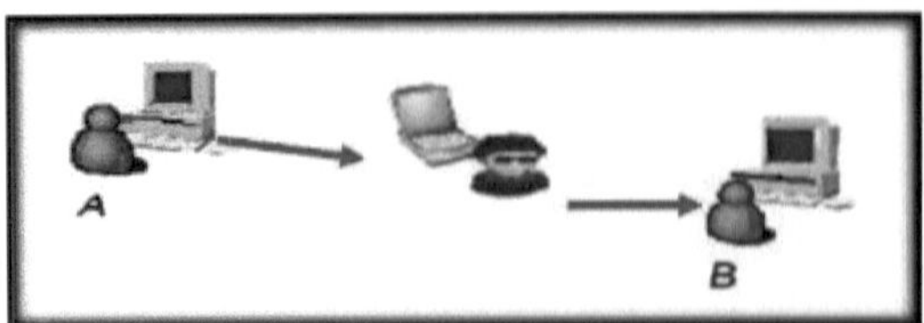

Figure A.7: Modification

> *Usurpation*

The attacker constructs the information stream himself, and sends it to the destination posing as the legitimate source.

Figure A.8: Spoofing

Frequent attacks

> ***Dictionary attacks***

The dictionary attack is a method used in cryptanalysis to find a password or clë. It involves testing a series of potential passwords, one after the other, hoping that the password used for encryption is contained in the dictionary. If this is not the case, the attack will fail.

This mëthod is based on the fact that many people use common passwords (e.g. a first name, a colour, the name of an animal,etc). To protëger the password from this kind of attack a simple and effective practice is used leet (It is a ëenIIIre using alphanumëric ASCII ciracteres)

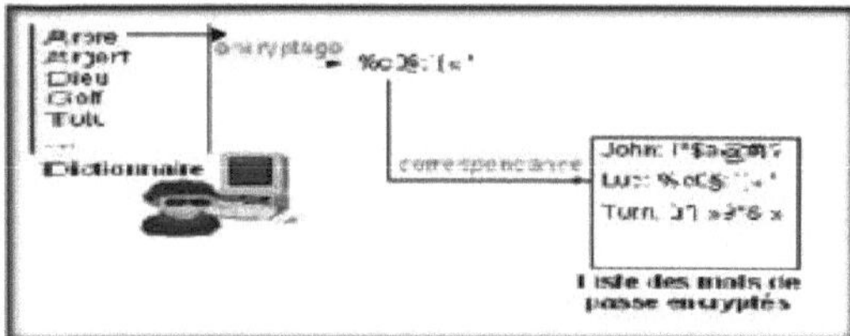

Figure A.9: Dictionary attacks

> ***Denis de service***

A denial of service attack is a type of attack aimed at making an organisation's services or resources unavailable for an indeterminate period of time. It is most often an attack on an organisation's servers, so that they cannot be used and accessed.

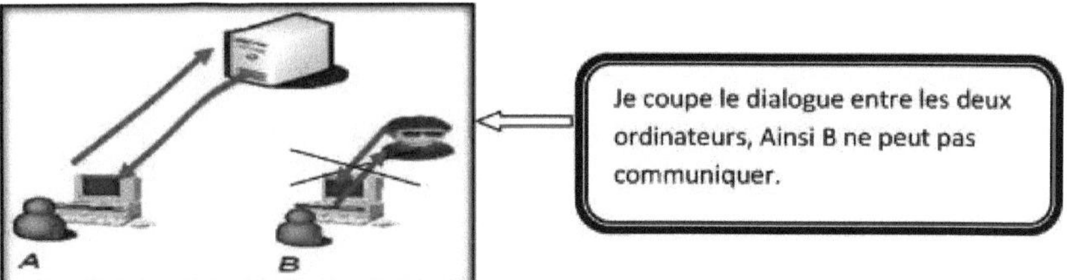

Figure A.10: Deni service

The principle of denial of service attacks consists of sending IP packets or data of unusual size or constitution, in order to cause saturation or an unstable state of the victim machines and thus prevent them from providing the network services they offer

> ***Man in the middle***

The main objective of this attack is to hijack the traffic between two machines. This is to intercept, modify or destroy the data transmitted during the communication. This attack is more a concept than a full-blown attack.

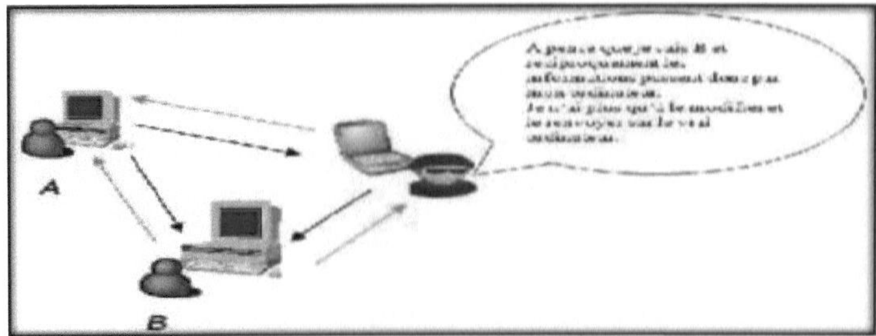

Figure A.11: Home in the middle

> ***Sniffing***

Sniffing is a technique that consists of analysing network traffic. When two computers communicate with each other, there is an exchange of information (traffic). They can then analyse it and find sensitive information. To protect yourself, you can use a switch instead of a hub, use encrypted protocols for sensitive information such as passwords and use a sniffer detector.

Figure A.12: Sniffing

> ***Scanning***

Scanning consists of scanning all the ports on a machine using a tool called a scanner. The scanner sends packets to several ports on the machine. Based on their reactions, the scanner will deduce if the ports are open (reply). This is a very useful tool for hackers. It allows them to know the weak points of a machine and thus to know where they can attack. Today, they can determine the operating system and applications associated with the ports. To protect yourself, you can scan the machine to find out which ports are open, monitor the open ports with a firewall, close the ones that are not useful and use an IDS (Intrusion Detection System) or better still an IPS (Intrusion Prevention System).

Figure A.13: Scanning

> ***Viruses***

A virus is a program that replicates itself by partially inserting itself into other files. Some viruses will simply show up as a message on your computer, others will be dangerous. They can delete data or format a hard disk. Most viruses are spread by e-mail attachments.

> ***To***

A worm is a program that spreads from computer to computer via a network such as the Internet. Thus, unlike a virus, the worm does not need a host program to ensure its reproduction. Its size is very small, which allows it to spread at an impressive speed over a network and use network resources, possibly bringing the network to saturation.

The worm can therefore not only affect a computer, but also degrade network performance. Like a virus, the worm can contain a harmful action that can be very serious such as formatting your hard drive or sending confidential data.

Spy

Spyware is harmful software that transmits information from your computer to third parties. Spyware is often found in freeware (open source software) or shareware (proprietary software). In general, open source software such as Mozilla FireFox does not contain any spyware. The risk is the diffusion of personal data, which are then used by advertising companies. For example, the spy can send a list of visited sites and e-mail addresses, which can then be used to target interests. The presence of a large number of spies on a computer causes a decrease in performance.

More rarely it is the action of :

> ***Computer hacker***

During an attack, the hacker always keeps in mind the risk of being detected, which is why hackers usually favour rebound attacks (as opposed to direct attacks), which consist in attacking a machine through another machine, in order to hide the traces allowing to trace him (such as his IP address) and in order to use the resources of the rebound machine.

This shows the importance of protecting one's network or personal computer, it is possible to find oneself an "accomplice" in an attack and in the event of a complaint from the victim, the first person questioned will be the owner of the machine used as a rebound.

On the other hand, professional hackers are able to effectively use sophisticated automated tools to carry out their activities. Whether motivated by financial gain or the challenge of breaking through barriers, hacker attacks are more numerous and sophisticated and appear to be better coordinated. Hacker groups are likely to be better organised, more skilled and far more active than terrorist groups in exploiting critical information infrastructure vulnerabilities.

The result of an attack is :

- Either a disclosure of the information: violation of the confidentiality of the object;
- Either a modification of the objects: violation of the integrity of the object;
- Either a denial of service: violation of availability.

Furthermore, attacks can be direct or indirect.

A direct attack is aimed directly at the object.

In an indirect attack, information is, for example, retrieved from an object without attacking it. In this case of attack, it is possible to derive confidential information from an object without any access to that object, for example by using statistics. Database systems are subject to this type of attack where it is possible to query databases about an object.

Printed by Books on Demand GmbH, Norderstedt / Germany